ROBOTICS

Julia Wall

CONTENTS

THE BOT PACK

Have you ever dreamed of having a machine to do your chores? With robotics that dream could come true. Robotics is the branch of science that deals with making and using robots.

Robots are machines that make decisions and do work. Before robots, people had to do the work that animals and machines couldn't do. The first robot started work in 1954 in a grocery warehouse in South Carolina. Early versions of robots could only carry things around factories and do simple jobs. Now robots can do much more complex tasks. Step into the world of robotics!

People who work with robots call them "bots."

ROBOTIC SYSTEMS

All robots are machines, but not all machines are robots. To be a robot, a machine must be able to do three things.

- It has to be able to detect something using **sensors**.
- It has to be able to make decisions using a computer to "think."
- It has to be able to do something based on those decisions, such as solve a problem.

Toasters can toast bread, but they can't detect the bread, "think" about the best way to toast the bread, and then toast the bread in that way. Only if a toaster could detect, make decisions, and then do something about them would it be a robot. Therefore, toasters are not robots.

But this machine is a robot. It can detect things, make decisions, and do something about its decisions.

How Robotic Systems Work

Robots have four main parts.

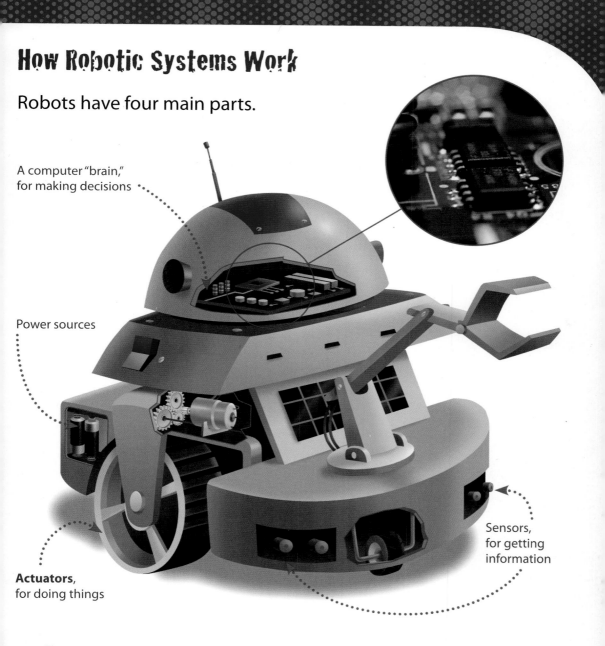

A computer "brain," for making decisions

Power sources

Actuators, for doing things

Sensors, for getting information

We control robots by programming their computers—putting instructions in their computers that tell them how to think. A robot that can learn from its mistakes has artificial intelligence (AI).

Algorithms

Algorithms are step-by-step rules for doing something. The programs that control robots use algorithms so that the robot knows what to do. Here is part of an algorithm a robot could use.

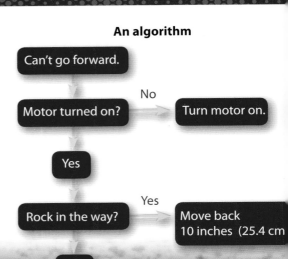

An algorithm

Can't go forward.

Motor turned on? — No → Turn motor on.

Yes

Rock in the way? — Yes → Move back 10 inches (25.4 cm

No

ROBOTIC APPLICATIONS

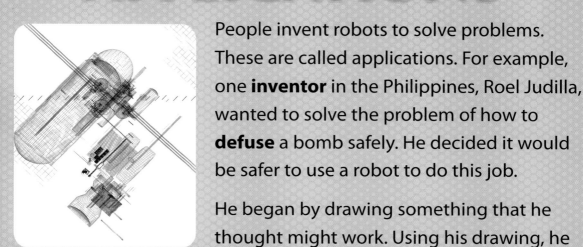

People invent robots to solve problems. These are called applications. For example, one **inventor** in the Philippines, Roel Judilla, wanted to solve the problem of how to **defuse** a bomb safely. He decided it would be safer to use a robot to do this job.

He began by drawing something that he thought might work. Using his drawing, he made a model.

Inventors make models from their drawings

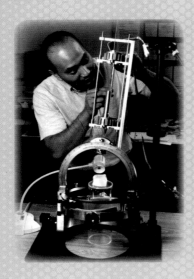

Inventors use models to search for problems, which **engineers** call "bugs." As they find each bug, they decide how to get rid of it.

Engineers like Judilla build robots to do jobs that are too dangerous for humans to do. Other engineers build robots that save people time.

This robot could save lives.

Medibots

Did you know that surgeons can operate on you without touching you? They use remote-controlled machines—machines they control from a distance.

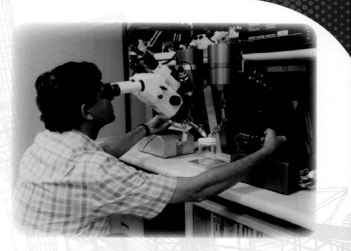

It's like playing a video game, only this is for real!

In surgery, medibots are the next step after remote-controlled machines.

medical + ro**bot** = **medibot**

Medibots will be machines that will work on their own, once a person has programmed their computer brain. They will be able to use very tiny instruments. This will make some operations easier.

Did you know that a baby's veins are only as thick as the hairs on your head? Imagine operating on something that tiny!

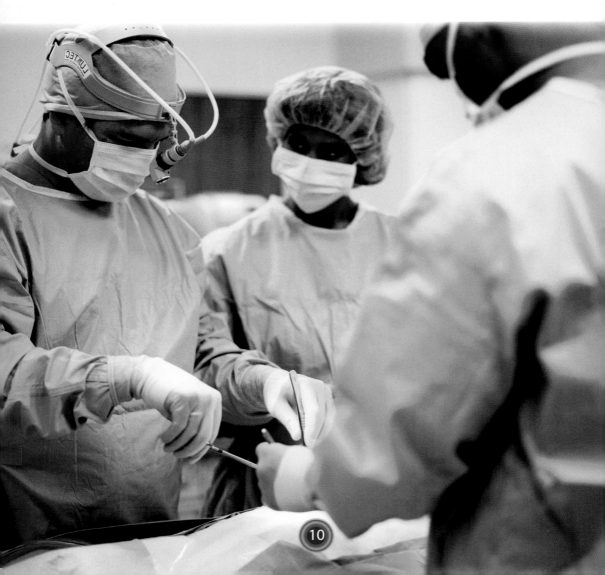

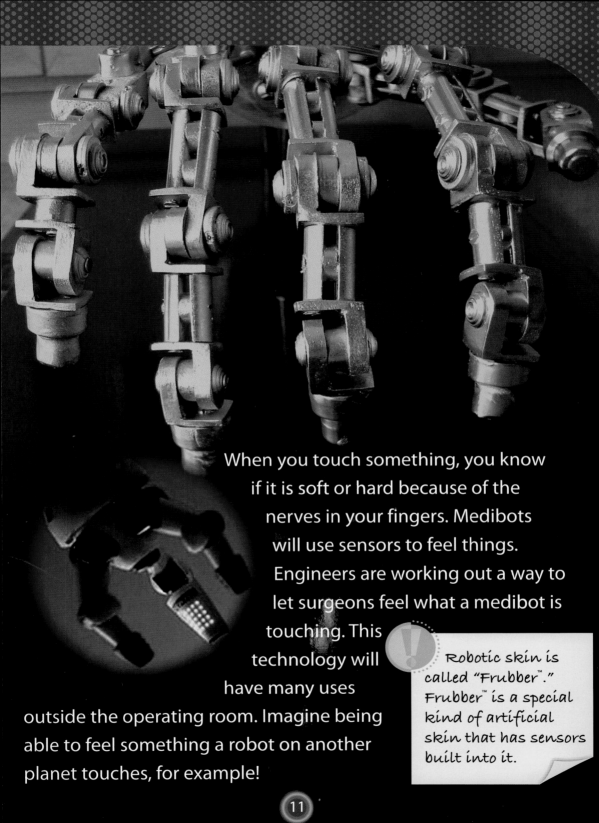

When you touch something, you know if it is soft or hard because of the nerves in your fingers. Medibots will use sensors to feel things. Engineers are working out a way to let surgeons feel what a medibot is touching. This technology will have many uses outside the operating room. Imagine being able to feel something a robot on another planet touches, for example!

Robotic skin is called "Frubber™." Frubber™ is a special kind of artificial skin that has sensors built into it.

Space Bots

We are already using space bots to explore outer space. Space bots are programmed to do things like **analyze** gas or pieces of rock and take photographs. Space bots are controlled by signals from earth when they are near earth. But what happens once they are a long way away?

Voyager 1 and *Voyager 2* explored the outer planets in our solar system. Then they became the first space bots to leave our solar system.

Once space bots are too far away for signals to reach them quickly, they need to have AI—they need to be programmed to do some things on their own.

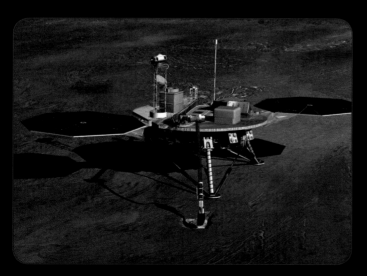

The Phoenix Mars Scout digs into the soil to see if there is ice underneath.

Cassini was the first space bot to **orbit** Saturn. It dropped another space bot onto Titan, one of Saturn's moons. Both bots sent information about Titan and Saturn back to earth.

Messenger is programmed to photograph Mercury's surface.

It takes 6 hours for a radio signal from earth to reach a space bot near Pluto. That's too long for a space bot in trouble!

Deep-Sea Explorers

About 60 percent of the ocean is more than 300 feet (91.44 meters) deep. The deepest parts are 7 miles (11.26 kilometers) below the surface! It's very dangerous for people to go down there, even in a submarine. The water pressure is enormous, and the water is extremely cold. Much of the ocean floor remains unexplored. Robots are changing this.

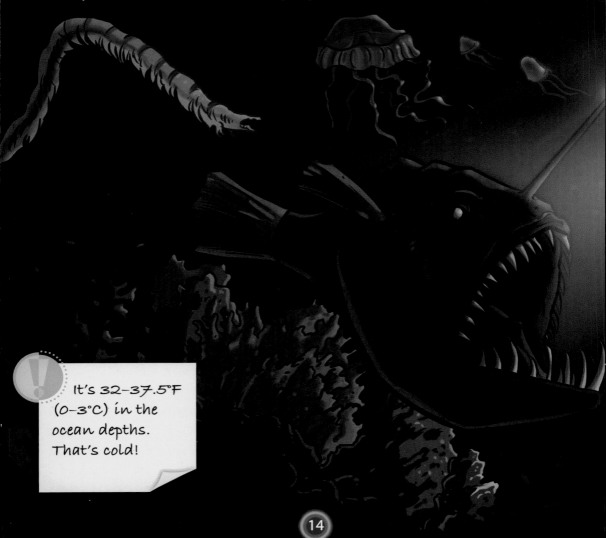

It's 32–37.5°F (0–3°C) in the ocean depths. That's cold!

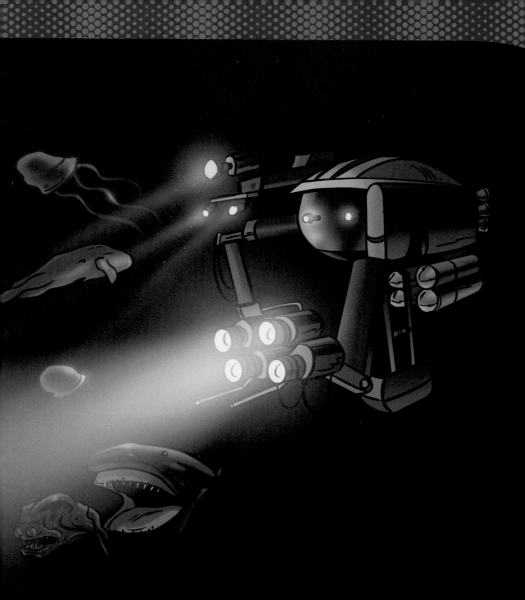

Some deep-sea bots are remote-controlled machines. Others are like space bots with AI and are programmed to do many things on their own.

Packbots

Packbots use cameras to see around corners and into tight spaces. They go where it's too risky for people to go, like into collapsed buildings after earthquakes.

Packbots that are programmed to make decisions, such as where to look next, are true robots.

They're called "packbots" because you can carry one in a pack on your back!

Nanobots

Nanobots are robots of the future. "Nano" means one billionth. Nanobots will be so small we won't be able to see them with ordinary microscopes. Scientists hope to make them as small as a billionth of a meter. Even though they will be incredibly tiny, they will still be able to measure things, "think," and do things.

1 meter = 3.28 feet. A billionth of a meter is about how far a fingernail grows in 1 second.

Imagine making robots this tiny. How will scientists put the parts together? Maybe they will need to build them one atom at a time. Maybe they'll use beams of light as tools to make them. Scientists don't know how to make nanobots yet, but they are figuring out how to.

In a way, bacteria are like naturally occurring, wild nanobots. However, bacteria can make you sick. Nanobots will make you better.

How will such tiny machines be powered? Most likely, nanobots will be powered by transducers. Transducers are electronic devices that change energy from one form into another. For example, nanobots might be able to use the energy in radio waves.

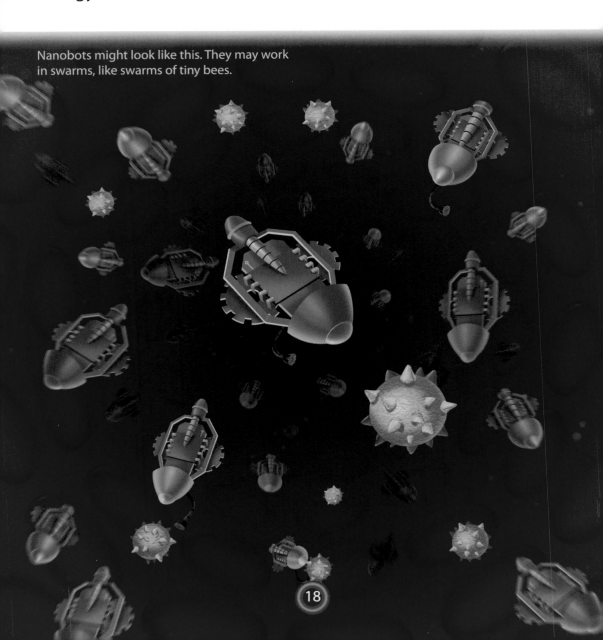

Nanobots might look like this. They may work in swarms, like swarms of tiny bees.

DESIGN AND BUILD A ROBOT

Did you know that you can design and build your own robot right now?

Design a Robot

You'll need to do some planning first. Here are some things to think about.

1. What do you want your robot to do? Do you have a problem a robot could solve? What about feeding a pet tarantula, for example?

2. What will your robot need to be able to do in order to carry out its task? If you want your robot to feed the tarantula a mealworm, it will need a camera so that it can see. It will need to be able to measure how far away the mealworm is. It will need a "hand" to pick up the mealworm.

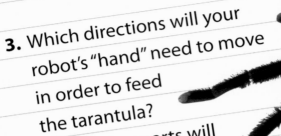

3. Which directions will your robot's "hand" need to move in order to feed the tarantula?

4. What other parts will your robot need? A programmable computer for its brain for sure, but what else?

5. How will your robot move around inside the tarantula's cage?

6. How will your robot be powered? What kind of transducer could you use, for example?

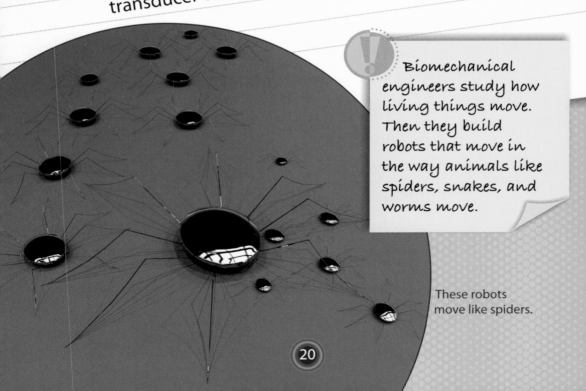

Biomechanical engineers study how living things move. Then they build robots that move in the way animals like spiders, snakes, and worms move.

These robots move like spiders.

Build a Robot

If you'd like to build a robot, you'll need some help.

Find a friend, an adult, or a group in your community to help you plan and build your robot. Some schools have clubs where kids build robots together.

Even if you don't join a club, you could still build and program a robot with help. You'll need computer **software** and machine parts from modeling stores, such as sensors and motors.

WILL ROBOTS "WAKE UP"?

When we think, **electrochemical** signals move through our brains at 492 feet per second. Computer "thought" moves at almost 900 million feet per second. This is millions of times faster than signals move in human brains.

We have invented robots. What if we go a step further and build robots that not only "think" faster than us but are also smarter?

Up to now, we've just been making better and better machines, but could artificially intelligent robots "wake up"? Robots that think for themselves open up a huge number of new possibilities but there could be unintended consequences we haven't thought of. The future is exciting, but we need to be careful.

GLOSSARY

actuators—the parts of a robot that carry out tasks

analyze—measure and study

defuse—take away the part of a bomb that makes it blow up

electrochemical—describes electricity produced by a chemical change

engineers—people who design, build, or maintain structures and machines, such as robots

inventor—the first person to figure out how to make something

orbit—circle around a moon, planet, or sun

sensors—mechanical or electronic parts that can do things like taste, smell, see, hear, and feel

software—the programs computers use